OFF-GRID SOLAR POWER

"Manual Book For Beginners To Create And Installed
Solar Power Circuit And Basic Of Electricity
Circuit Ampere, Voltage, Ohm"

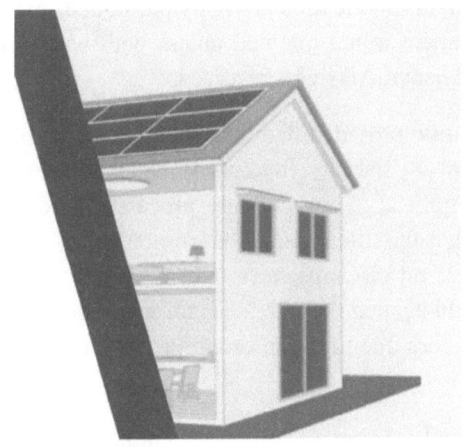

FWG Franklin White Group

TABLE OF CONTENTS

INTRODUCTION

Off-grid solar power. When you read these words, what first comes to mind? Is it something like, "can electricity be free?" or "nothing is off-grid?" When we talk about off-grid, what is a grid itself? A grid is a system of parallel crossing lines, regardless of whether genuine or fanciful. Most American lanes are spread out in a grid design, which means the paths cross at the right points to create squares when seen from above.

You've presumably observed a map grid, the uniform lines drawn on a map that permits you to pinpoint a specific area. The grid can likewise allude to a physical system of sorts, not made of straight or parallel lines. You might be comfortable with the high voltage electrical links that convey power through the nation, known as the national grid. A grid can likewise be a gadget comprised of crossing metal bars that you use when barbecuing food.

Off-the-grid, or off-grid, is an attribute of structures and a way of life planned on a freeway without dependence on at least one open utilities. The expression "off-the-grid" generally alludes to not being associated with the electrical grid, yet can likewise incorporate different services like water, gas, and sewer frameworks, and can scale from private homes to little networks. Off-the-grid living requires structures and individuals to act naturally adequate, which is beneficial in disconnected areas where standard utilities can't reach and are appealing to the individuals who feel the need to decrease their ecological effect and average cost for necessary items. By and large, an off-grid building must have the option to flexibly vitality and consumable water for itself, just as it oversees food, waste, and wastewater.

Weill is relating to the history of electric energy. We will be focusing on solar because solar energy is a great typical version of an off-grid electric supply for homes. I will also teach the steps to arranging the solar panel and installing it.

CHAPTER 1: ENERGY

WHAT DO YOU KNOW ABOUT "ENERGY"?

Before we can talk about how solar panels work, we need to talk about energy itself. Many people have done many things with energy generation, such as taking energy from the sun and converting it into electric energy. Energy is a principal physical amount that assumes a focal job in physics and innovation, science, and the economy. Its SI unit is the Joule. The practical meaning of energy is often that a physical system can give off heat, work, or emit radiation to the extent that its energy is reduced. In a system that is closed to the environment, the total energy does not change. The importance of energy in theoretical physics is, among other things, that the principle of energy conservation—orignially a fact of experience—can be deduced from the fundamental physical laws of nature that cannot be changed over time.

Energy is available in different forms of energy that can be converted into each other. Examples of energy forms are potential, kinetic, electrical, chemical, and thermal. Examples of such conversions of energy are a person lifting a parcel or accelerating a bicycle, charging a battery, the body metabolising, or the heat given off by a heater.

The energy of a system depends on its state, e.g. the parameters of the system and the current values of its variables. According to Hamiltonian equations of motion, the Schrödinger equation, or the Dirac equation, the form of this dependency determines the development of the system over time in every detail.

As per the hypothesis of relativity, rest vitality and mass are controlled by the proportionality of mass and vitality associated.

The word vitality returns to old Greek ἐνέργεια, energeia, which in Greek culture had simply a philosophical significance in the feeling of "living reality and viability."As a logical term, the word itself was first brought into mechanics by the physicist Thomas Young in 1807. The new amount of vitality ought to demonstrate the quality of specific impacts a moving body can deliver through its development, not just through its drive ("Mass occasions speed"). The impulse has been over since Christiaan Huygens, Christopher Wren, and John Wallis examined two bodies' colliding around 1668. It was known that it remains with elastic and inelastic bodies, so it is the right measure for the changes caused and thus for the permanent "size of the movement." However, in other processes, bodies of different masses, even if they have the same impulse, cause effects of different sizes. These include, for example, the height that a body reaches in an upward movement or the depth of the hole that it hits in a soft mass on impact. The effect does not increase proportionally with speed like the impulse, but with the square of the speed. Therefore, Gottfried Wilhelm Leibniz designated the size in 1686 display style mv as the true measure of the movement's size and called it vis viva ("living force"). This name followed the language used at the time, in which a body could only produce effects through the forces inherent in it. The name living force but by "confusion with the Newtonian concept of force a disastrous confusion of ideas and a countless multitude of misunderstandings caused" (as Max Planck in 1887 in his award-winning portrayal of the history of energy conservation) argued Leibniz as follows:

weight of to the height lifting takes as much work as a weight to the height to lift (lever law). According to Galileo, Galilei is in free fall, so the first case's top speed is twice as high as in the second case. The inherent (living) power with which one wants to measure this work (a latent form of the living force) is with preservation of the living force, not as the followers of Descartes thought.

The correct pre-factor Daniel Bernoulli derived in kinetic energy in 1726. As found by other analytical mechanics of the 18th century, such as Leonhard Euler and Joseph Louis Lagrange, there are precursors to the concept of potential energy. The term "potential function" originated with George Green in 1828 and was independently introduced by Carl Friedrich Gauss in 1840 (but was already known as a potential Lagrange and Laplace). The concept was already known to Leibniz in its derivation form and Johann I Bernoulli, who was the first to formulate the principle of preserving the living forces in 1735. Still, Leibniz also had the idea, for example, in the 5th letter to Samuel Clarke, which was particularly popularised by a student of Leibniz, Christian Wolff. At that time, one spoke of potential energy as the latent form of the living force, which, for example, is distributed to smaller particles of the body during the inelastic impact.

To predict the effects of movement of the body, Young defined energy as the body's ability to travel a certain distance against a resisting force. He also noticed that work done in the form of lifting later found itself quantitatively in its energy. Additionally, Young did not come up with the concept of converting different forms of energy, and he kept the Leibniz formula but, on the whole, was still a follower of the Cartesian view of the forces.

In the 18th century, mechanics and physics were not particularly interested in energy. Important researchers, such as Euler, saw the dispute over the vis viva, the true measure of force, as a matter for the philosophers and instead dealt with the equations of motion, especially in celestial mechanics. The concept of energy in today's sense did not originate from the analytical mechanics of the 18th century, but from the applied mathematicians of the French school, including Lazare Carnot, who wrote that the living force is either as or can manifest force times away (as latent living force). A quantitative definition of the work ("force times way" was also given simultaneously by Coriolis and Poncelet in 1829, apparently independently of one another and Young). Coriolis also found the right expression for the kinetic energy, which Rankine first called kinetic energy in 1853.

Regarding the steam motor, the thought built up that heat vitality is the reason for moving vitality or mechanical work in numerous procedures. The beginning stage was that water is changed over into the vaporous state by heat, and the gas extension is utilised to move a cylinder in a chamber. The forced movement of the piston reduces the stored thermal energy of the water vapour. The connection between mechanical energy and heat was demonstrated in experiments famous by Benjamin Thompson (Graf Rumford, Munich 1796, 1798) and Humphry Davy (1799).

The physicist Nicolas Carnot recognised that a change in the steam volume is necessary when performing mechanical work. He also found out that cooling the hot water in the steam engine is not just a matter of heat conduction. Carnot published these findings in 1824 in a highly-regarded document on the steam engine's principle of operation. Émile Clapeyron converted Carnot's findings into a mathematical form in 1834 and developed the Carnot cycle's graphic representation that is still used today.

In 1841, the German specialist Julius Robert Mayer distributed his thought that vitality can not be made nor demolished; however, it must be changed over. He kept in touch with a companion: "My case is: Falling power, development, heat, light, power, and the concoction distinction between the ponderability are very much the same items in various appearances." The measure of heat that is lost in a steam motor compares precisely to the mechanical work done by the machine. This is referred to today as "vitality preservation," or "first law of thermodynamics." The physicist Rudolf Clausius improved the ideas about energy conversion in 1854. He showed that only part of the thermal energy could be converted into mechanical work. A body where the temperature stays steady can't accomplish mechanical work. Clausius built up the second law of thermodynamics and presented the idea of entropy. As indicated by the subsequent law, it is outlandish for heat to go from a colder to a hotter body autonomously.

Hermann von Helmholtz formulated the principle "about the maintenance of strength," and the Perpetuum mobile the 1st kind in 1847 generated energy when inserted. Von Helmholtz discovered his insight by working with electrical vitality from galvanic components, particularly a zinc/bromine cell. In later years, he joined the entropy and the heat improvement of a compound transformation to free vitality. Nonetheless, both Mayer and von Helmholtz experienced issues distributing their discoveries during the 1840s, since both were first considered untouchables. The physicists in Germany were in a protective position against the characteristic way of thinking of Schelling's region, which had been persuasive since the finish of the eighteenth century. Both associated with being supporters of this theoretical material science. Josiah Gibbs came to similar knowledge as von Helmholtz in 1878 regarding electrochemical cells. Chemical reactions only take place when the free energy decreases. Free energy can predict whether a chemical conversion is possible or how the chemical equilibrium of a reaction behaves when the temperature changes.

After Wilhelm Wien (1900), Max Abraham (1902), and Hendrik Lorentz (1904) had already published considerations on electromagnetic mass, Albert Einstein published the finding in 1905 as part of his special theory of relativity that mass and energy are equivalent.

FORMS OF ENERGY AND ENERGY CONVERSION

Energy can be contained in a framework in various ways. These possibilities are called forms of energy, such as:

- Kinetic energy

- Chemical energy
- Electrical or potential energy.

Different forms of energy can be converted into each other, the sum of the amounts of energy over the different forms of energy before and after the energy conversion always being the same.

A conversion can be carried out so that all other conservation parameters of the system will have the same rate of value before and after the conversion. For example, the conversion of kinetic energy is restricted by maintaining the system's momentum and angular momentum. A gyroscope can only be braked and thus lose energy if it simultaneously emits angular momentum. There are also limitations at the molecular level. Numerous possible responses don't happen because they would abuse energy protection. Other conservation parameters are the number of baryons and the number of leptons. They limit the conversion of energy through nuclear reactions. The energy contained in the mass of the issue must be changed over into another type of energy with an equivalent measure of antimatter.

Without antimatter, the change with atomic fission or fusion assistance just attempts to a small degree. The thermodynamics is the second law of thermodynamics, another condition for conversion before a closed system's entropy cannot decrease. They are extracting heat without other processes, running in parallel means to cool down. Be that as it may, a lower temperature relates to decreased entropy and, in this manner, repudiates the subsequent law. Another piece of the framework must be heated in kind for cooling to change heat into another type of energy. The transformation of heat energy into different types of energy in this manner consistently requires a temperature contrast. Also, the aggregate sum of heat put away in the temperature contrast can't be changed over. Heat motors are utilised to change heat into mechanical energy. The proportion of the most extreme conceivable work constantly allowing the law to measure heat expended is called Carnot productivity. The more noteworthy the temperature distinction with which the heat motor works, the more prominent it is. Other conversions are less affected by the constraints of conservation laws and thermodynamics. In this way, electrical energy can be converted almost completely into many other energy forms with little technical effort. For example, they convert electric motors into kinetic energy.

Most conversions do not take place entirely in a single form of energy, but part of the energy is converted into heat. In mechanical applications, the heat is mostly generated by friction. In electrical applications, electrical resistance or eddy currents are often the cause of the generation of heat. This heat is usually not used and constitutes loss. In connection with electrical current, the emission of electromagnetic waves can also occur as an undesirable loss. The relationship between successfully converted energy and energy used is called efficiency.

Many energy conversions are often coupled in technical applications. In a coal-terminated force plant, the coal's substantial energy is first changed over into heat by burning and moved to water fume. Turbines convert the steam's warmth into mechanical energy and thus drive generators that convert the mechanical energy into electrical energy.

The energy in classical mechanics

The pendulum of a clock regularly converts kinetic energy into potential energy and vice versa. The watch uses the weights' position and energy in the earth's gravitational field to compensate for frictional losses.

In old-style mechanics, the energy of a framework is its capacity to work to do. The work converts energy between different forms. Newton's laws states that the sum of all energies does not change. Friction and the associated energy losses are not taken into account in this analysis.

The Noether Theorem allows a more general definition of energy which automatically considers the aspect of energy conservation. Every single normal law of old-style mechanics is invariant as to shifts in time. They are described by the way that they apply in a similar structure unaltered consistently. The Noether theorem now states that there is a physical quantity for this symmetry concerning displacement in time, the rate of value of which does not change with time. What size is the energy?

Due to erosion, it adheres to from the energy protection law and inescapable energy misfortunes that it is difficult to construct a mechanical machine with sudden spikes in demand for its own for whatever ideal length of time (Perpetuum versatile). Also, together with the conservation of momentum, the conservation of energy provides an explanation for collisions between objects, without having to know the exact mechanism of the collision.

ENERGY AND MOVEMENT

Motor energy is energy required for movement of the body. It corresponds to the mass and the square of the speed comparative with the inertial frame in which you depict the body. Therefore, the amount of kinetic energy depends on the point of view from which the system is described. An inertial system is often used, which rests on the ground.

Notwithstanding a translational development, a body can likewise play out a revolving development. The active energy contained in the turning movement is called rotational energy. This corresponds to the square of the rakish speed and the snapshot of the body's latency.

ELECTRIC ENERGY OR POTENTIAL

Electric energy or potential energy is caused when a body is in a force field for, as long as it is a conservative force act. This could be the earth's gravity field or the force field of a spring, for example. The potential energy diminishes toward the power and increments against the course of the power. Opposite to the bearing of the power, it is consistent. If the body moves from high potential energy to less potential energy, it does just as much physical work how his potential energy has decreased. This statement applies regardless of how the body got from high potential energy to less.

The expected energy of a body in a homogeneous gravitational field with increasing gravitational speed is relative to the height over the arrange framework's root:

In free fall, this potential energy is changed over into kinetic energy by quickening the body.

Since the organising starting point can be picked subjectively, the body's positional energy is rarely completely given and not quantifiable. Only their progressions are quantifiable. Periodic movements regularly convert potential into kinetic energy and back into potential energy. With a pendulum, the potential energy is at a maximum at the reversal points; the kinetic energy is zero here. When the thread is hanging vertically, the mass reaches its maximum speed and, thus, also its maximum kinetic energy; the potential energy has a minimum here. A planet has the highest potential and the lowest kinetic energy at its point furthest from the sun. Up to a point closest to the sun, its orbital speed increases just enough that the increase in kinetic energy exactly compensates for the decrease in potential energy.

Flexible energy is the possible energy of the iotas or atoms dislodged from their rest position in a flexibly twisted body, such as mechanical spring. The energy stored (or released) in the body during elastic or plastic deformation is generally referred to as deformation energy.

ENERGY IN THERMODYNAMICS

Warm energy is stored in the scattered development of the particles or atoms of a substance. Informally, it is likewise alluded to as "heat energy" or "warmth content." The transformation of warm energy into different types of energy is depicted by thermodynamics. The framework (interior energy, enthalpy) and the warmth, the thermal energy, is transported across the framework limit. The sum of thermal energy, vibrational energy in the body, and binding energy is called internal energy. Some sources also distinguish between internal thermal energy, internal chemical energy, and nuclear energy as internal energy, but this goes beyond the scope of thermodynamics.

Converting thermal energy into mechanical work

While all forms of energy can be completely converted into thermal energy under certain conditions stated by the first law of thermodynamics, this does not apply in the opposite direction. The second law of thermodynamics describes a very important limitation. Depending on the temperature, only a relatively large proportion can be converted into mechanical work via a cyclical process. At the same time, the rest is released into the environment. In technical thermodynamics, the convertible parts of an energy form are also called exergy. The exergy is certainly not a state variable in the genuine sense, since it depends on the condition of the framework in addition to the condition of the earth, which is given in the individual case. Exergy stream outlines of an energy transformation chain could then be utilised to follow where avoidable misfortunes (erosion or other dissipative procedures) can be recorded. You can see that when chemical energy (100% exergy) is converted into heat at an average temperature of 1000 ° C, the exergy share is only 80%. This energy is used as heat in a steam boiler to steam transferred at 273 ° C, only about 50% remain and only about 7% when transferred to a room heated at 20 ° C.

CALCULATION OF THE MAXIMUM WORK (EXERGY)

To calculate the exergetic proportion of thermal energy, consider whether the heat source has a constant temperature, such as in a boiling water reactor which stays at approximately 270 ° C, whether or not the heat output of its cooling medium flue gas takes place. The exergetic component can be determined via The relationship can likewise be seen on the charts: T the outright temperature in K; S the entropy in J/K; H the enthalpy in J file 1: beginning state list; and U: surrounding state.

The enthalpy distinction is basically (in this situation) the energy provided as the heat from the burning air-fuel. It shows up as a territory under the isobaric heat gracefully bend. The exergetic part is over the encompassing temperature, the other non-usable part, called "energy," is underneath this line. The reduction in exergy in an energy transformation bind is likewise alluded to as energy downgrading.

When the heat is moved from the vent gas to the working medium, the water that vanishes and overheats, there is a further loss of exergy. For a procedure with superheated steam of 16 bar and 350 ° C, for instance, the greatest mechanical force that can be gotten from the steam mass stream should never be determined to utilise the Carnot productivity at this temperature. The outcome with an effectiveness of 52% would not be right. It would negate the second law because the mean temperature of the heat contribution to the water-steam cycle is lower. There is no inward heat move from gathering steam to the feed water, for example, in steam motors. In the best case hypothetically, the steam can be reversibly brought to water with encompassing conditions. The greatest effectiveness of 34.4% is accomplished at a surrounding temperature of 15 ° C. Interestingly, the reversible Clausius-Rankine process in Figure 4 with a steam weight of 32 bar and buildup at 24 ° C arrives at 37.2%. With these steam boundaries, genuine procedures just accomplish far lower efficiencies.

CHAPTER 2: TYPES OF ENERGY

SOLAR ENERGY

The solar energy that reaches the earth through radiation also experiences an exergy loss on the way to the earth's surface. While the sun's inner energy at around 15 million K is practically pure exergy, the sun shines on the earth's surface at a surface temperature of around 6000 K, the temperature of which is around 300 K. By concentrating the sun's rays in a collector, you wouldn't get beyond the sun's surface temperature even in the high mountains, where absorption by the earth's atmosphere hardly plays a role. An efficiency would result in the Carnot factor of about 95%. Then, however, no more energy would be transmitted. The thermodynamic limit is below an absorber temperature of 2500 K with an efficiency of approximately 85%. In practice, there are dissipative losses, from absorption in the atmosphere through the crystalline cells' material properties to the ohmic obstruction of the photovoltaic frameworks. To date, only efficiencies of less than 20% can be achieved. The highest efficiency currently achieved is 18.7%.

COMBINED HEAT AND POWER (CHP)

Heating with only a small amount of exergy is usually required for heating. That is why heating with electrical current via resistance heating is "energy wasted." Any place mechanical energy or power is produced from heat, there is a requirement for heat. The utilisation of waste heat bodes better than alternative arrangements of heat. When steam is utilised in a heat force station, steam is expelled from the turbine, the temperature of which is sufficiently high to lead the buildup heat to the buyer using a local heating system. Alternatively, the waste heat from stationary combustion engines is also used in combined heat and power plants. The heat pump uses work to absorb heat (energy) from the environment and to release it together with the drive work as heating energy at a correspondingly high temperature. If groundwater at 10 ° C is available as a heat source and a room should be warmed at 20 ° C, a warm siphon with the Carnot process could deliver 29 kWh of heat by using a one-kilowatt hour of drive work (work number = 29). Real heat pumps, which are operated by alternately evaporating and condensing refrigerants at different pressures, achieve approximately 3 to 5 (working figures).

CHEMICAL ENERGY

Chemical energy is the form of energy stored in the form of a chemical compound in an energy carrier and can be released during chemical reactions. It describes the energy associated with electrical forces in atoms and molecules, and it can be divided into the kinetic energy of the electrons in the atoms and the electrical energy of electrons and protons' collaboration.

ENERGY IN ELECTRODYNAMICS

If there is no magnetic field that changes over time, an electrical potential can be defined in an electrical field. At that point, a charge transporter has possible electrical (electrostatic) energy that corresponds to the potential and its measure of charge. Since the zero purposes of the potential can be unreservedly decided, the energy is likewise not characterised. For two focuses in the likely field, in any case, the distinction in energies is autonomous of the decision of the expected zero points. Potential differences correspond to voltages in electrical engineering; the earth's capability is typically picked as the zero purposes of the possible scale. For arrangements of two electrical conductors, the electrostatic energy is proportional to the square of the difference in the two conductors' electrical potentials. Double the proportionality constant is called electrical capacitance. Capacitors are electrical engineering components that have a high capacity and can, therefore, store energy.

Equivalent to the view that charges carry electrostatic energy is the interpretation that the energy is distributed among the space between the charges. The energy density, i.e. the energy per volume element, is proportional to the electric field strength square. If there is a dielectric in the electrical field, the energy is also proportional to the dielectric constant.

If a charge moves in a vacuum to a place where the electrical potential is lower, the kinetic energy of the charge incrementally diminishes the same amount as the potential energy. For instance, this occurs with electrons in an electron tube, an X-bar tube, or a cathode shaft tube screen. Conversely, if a charge moves along a possible inclination in a conductor, it quickly discharges its consumed energy as warmth to the conductor medium. The force is relative to the expected inclination and the current.

Electrical energy can be shipped by moving charge transporters along conveyors with no noteworthy possible drop. This is the case, for example, in overhead lines or power cables, through which electrical energy flows from the power plant to the consumer.

Magnetic energy is contained in magnetic fields like in the superconducting magnetic energy storage.

Energy stored in an ideal electrical resonant circuit changes continuously between the electrical form and the magnetic form. The sum of the partial energies is the same at all times (energy conservation). The pure magnetic or electrical component of the energy has twice the frequency of the electrical oscillation.

ENERGY IN RELATIVITY

According to the special theory of relativity, corresponds to the mass resting energy of a resting object. Thus, the resting energy is down to the factor (square of the speed of light equivalent to the mass). The rest of the energy can be converted into other forms of energy in certain processes and vice versa. The reaction products of nuclear fission and fusion have measurably lower masses than the starting materials. In elementary particle physics, the generation of particles and thus of resting energy from other energy forms is observed vice versa.

In old-style mechanics, the rest energy isn't tallied, since it is irrelevant as long as particles do not convert into other particles.

The general theory of relativity further generalises the concept of energy and contains a uniform representation of energies and impulses as sources for spatial curvatures via the energy-impulse tensor. From this, the measurable quantities, such as energy density, can be obtained through contractions. The energy content is decisive for the study of the development of spacetime. In this way, one can predict the collapse of spacetime into a singularity from energy conditions.

THE ENERGY IN QUANTUM MECHANICS

In quantum mechanics, the Hamilton operator determines what energy can be measured on a physical system. Bound conditions of the framework can compare to discrete, not discretionary, energy esteems. In this way, the particles or beams radiated at advances between these states have line spectra. The quantisation of energy occurs with electromagnetic waves. A wave of frequency can only pack energy, which is the Planck quantum of action.

TECHNICAL USE OF ENERGY

The generation of energy is not possible due to the energy conservation rate. The term energy generation is nevertheless used in business life to express the conversion of a certain form of energy (e.g. electrical current) from another form (e.g. chemical energy in the form of coal). Similarly, there is no energy consumption in a strictly physical sense. Economically, it means the transition from usable primary energy, such as oil, gas, and coal to a form of energy that can no longer be used, for example, waste heat in the environment. From sparing, energy is utilised when progressively productive procedures are discovered that utilize less essential energy for a similar reason or something else. For instance, by previous utilisation, the essential energy utilisation is diminished.

Material science portrays the "energy utilisation" coolly presented above with the specific term entropy increment. While energy is constantly held in a shut framework, entropy consistently increments after some time or stays steady, best case scenario. The less usable the energy, the higher the entropy. As opposed to grow-ing entropy, one can, in a like manner, talk plainly of energy depreciation. Specifically, the law of entropy increment keeps heat energy from being legitimately changed over into kinetic or electrical energy. Rather, a heat source and a heat sink (= cooling) are constantly required. The most extreme productivity can be deter-mined from the temperature distinction as per Carnot.

The marginal instance of an energy transformation without an expansion in entropy is known as the re-versible procedure. A case of a practically reversible energy change is a satellite in a curved circle around the earth: at the most noteworthy purpose of the circle, it has likely high energy and low kinetic energy, and at the absolute bottom of the circle is the inverse. The change can happen a thousand times each year without noteworthy misfortunes. In superconducting resona-tors, energy can be changed between radiation energy and electrical energy millions or even billions of times each second, moreover with misfortunes short of what one for each thousand for every transformation. For many processes that were once associated with high losses due to significant increases in entropy, techno-logical progress enables increasingly lower losses. An energy-saving lamp or LED converts electrical energy into light much more efficiently than an incandescent lamp. A heat pump often generates much more heat than a conventional electric heater with the same output using heat from the environment for a certain electrical output. In other areas, however, state of the art technology has been close to the theoretical maximum for some time, so only small advances are possible. Good electric motors convert over 90 percent of the electrical energy used into usable mechanical energy and only a small part into useless heat.

Saving energy in the physical sense means minimising energy devaluation and the increase in entropy when converting or using energy.

SPECIFIC ENERGY

Specifically, in the natural sciences, it means related to a certain assessment basis. The particular energy is identified with a specific property of a framework, which can be described by a physical amount.

Examples

- Energy per volume in J/m^3: Enthalpy (thermodynamics); explicit inert heat: the heat of fusion; the heat of vaporisation; heat of crystalli-sation or the relating enthalpies (Material Science); gathering and heating worth (Energy Technology); explicit compaction energy (Material Science); explicit energy of hazardous.

- Energy per mass in J/kg (explicit measurement) work; explicit inert heat (thermodynamics); calorific worth and calorific estimation of strong energises; explicit energy of the energy stockpiling (energy innovation); electrical limit and energy thickness of the plate capacitor (electrical building); explicit energy of the mass point (mechanics)

- Thermodynamics and chemistry describe substance-related energy values not as specific, but as molar:

- Energy per measure of substance in J/mol (measurement) molar inactive heat (thermodynamics)

Energy supply and consumption

Colloquially, energy consumption is the use of different energies in forms that can be used by humans. Energy supply means the supply of consumers with these forms of energy, including the necessary energy infrastructure.

The most common forms of energy used by humans are thermal energy and electrical energy. Human needs are primarily focused on heating, food preparation, and the operation of facilities and machines to make life easier. On the subject of locomotion, the consumption of fossil energies as fuels for vehicles is important.

The various energy sources can reach the consumers via lines, such as fuel, electrical energy, and heating energy, or they are largely storable and easy to transport, such as coal, heating oil, other fuels (petrol, diesel, kerosene), nuclear fuel, and biomass.

Energy requirements are very different worldwide and many times higher in more industrialised countries than in developing countries. In industrialised countries, companies have been involved in the generation and supply of energy for general consumption since the 19th century. The focus today is on the central generation of electrical energy and distribution to the individual consumers. The procurement, transport, and refinement of fuels for heating purposes are also important economic sectors.

In addition to the derived SI unit Joule, other energy units are used depending on the application. Watt seconds (Ws) and Newton-meters (Nm) are identical to the Joule.

The electron volt (eV) is utilised in nuclear material science, atomic physical science, and rudimentary molecule material science to demonstrate molecule energies and energy levels. Rydberg is less basic in nuclear material science. The cgs unit erg is regularly utilised in hypothetical material science.

The calorie was used in calorimetry but is now only utilised informally and in exchange expansion to the legitimate unit joule while expressing the physiological calorific estimation of food. Energy providers measure the energy conveyed to clients in kilowatt-hours (kWh). The hard coal unit and the oil unit serve to indicate the energy content of primary energy sources. The TNT equivalent is used to measure the explosive power of explosives.

CHAPTER 3: RENEWABLE ENERGY

Renewable energies, also called regenerative powers, are energy sources that are infinitely available or can grow back in a shorter time - in contrast to fossil energy sources such as coal or natural gas. Renewable energy sources include hydropower, solar and wind energy, biomass, and geothermal energy. Renewable energies are an integral part of today's energy landscape. Until the beginning of this century, renewable energies were still models of power generation that could never completely lose their experimental character, but renewable technologies became increasingly important in the 00s. A fundamental social change has recently occurred due to the reactor disaster in Fukushima. This is reflected on the one hand in the phase-out of nuclear power in the Federal Republic and the EEG amendment of 2012, but can also be seen in broader social discussions and global changes, such as the 2015 World Climate Summit.

This change also has consequences for the German electricity landscape: the share of renewable energies has amplified intensely in the past 20 years. Distribution of gross renewable electricity consumption reached 42.1% in 2019, while the share was still 7.7% in 2002 and 3.4% in 1990. Whereas large coal and nuclear power plants took over a large part of the power supply in the past, this responsibility is now divided among a large number of smaller entities. Classic renewable energy systems range from a few kilowatts to several megawatts. However, they all produce electricity independently of one another. This is one of the major challenges of the energy transition: coordinating many decentralised electricity producers. This is where technological developments such as virtual power plants come into play, which, since the EEG 2012, have been helping to safely integrate the electricity production of many small plants into the German energy landscape. In this context, short-term electricity trading had also become much more important, since electricity production had become much more volatile than before when only a handful of power plants produced almost all of the electricity required.

Renewable energies are not only shouldering almost half of electricity production today. They also make an important contribution to system stability by providing system services such as balancing energy.

HISTORY OF RENEWABLE ENERGIES

PHOTOVOLTAIC

The first photovoltaic cells were used in 1958 on the mission of the US satellite Vanguard. However, it would take more than two decades before terrestrial systems were installed. In 1976, the Australian government decided to equip the telecommunications network in the outback with solar cells to charge the batteries installed there. Installations on oil rigs or the US Coast Guard in the 80s were the first, more widespread projects. In the mid-1980s, the Swiss engineer Markus Real installed small, decentralised PV systems on house roofs to demonstrate private implementation. Subsequently, numerous large-scale solar projects started, such as the 1,000-roof program in Germany (1990) or the 70,000-roof program in Japan (1994). In Germany, primarily small systems were initially installed, which also explains how in 2005, the total nominal PV power was only one gigawatt. In 2010 the 10 GW limit was exceeded in Germany, and in 2012, 25 GW were reached. At the end of 2019, over 49 GW were installed in Germany.

WIND POWER

As part of windmill technology, wind turbines were already in use in pre-industrial times—of course not as electricity-producing systems. The first attempts to generate electricity with wind turbines were made at the end of the 19th century. In the 1930s and 40s, the first successful tests with wind turbines were carried out in both the USA and Germany, but without regular use. The first plant that successfully fed in over a long time was the Gedser wind turbine in Denmark. In 1987 the first wind farm in Germany was built on the Growian site near Marne. Around 19 million kWh of electricity were produced there every year. In the course of the feed-in electricity law, the general expansion of wind turbines in Germany also grew during the 1990s. The boom was so great that two-thirds of European wind turbines were installed in Germany in the first half of the 00s. In 2019, the installed capacity of wind power (on- and off-shore) in Germany was 60.87 GW.

BIOMASS

In the early 20s, the first biogas plants were used, such as in the Ruhr area. However, these were wastewater treatment plants with fermenters. At that time, the gas produced there was not yet used to produce electricity, but was fed into the gas network. The first attempts to generate electricity from biogas were made in the 1930s and 50s, but this turned out to be uneconomical due to the high production costs. With the oil crisis, the perception of biogas in public discourse increased.

Nevertheless, biogas played a subordinate role until the late 1990s. Just under 700 systems were currently in operation. In the 00s, biogas became more and more important in the course of the first renewable energy laws and was even able to assume system responsibility from the EEG 2012. In 2012, the installed capacity of biomass plants reached 3000 MW. With the subsequent amendment to the EEG 2014; however, biogas was again assigned a smaller role in the energy mix, so that the number of new installations almost stagnated. Other energy sources from biomass, such as wood or waste-to-energy plants, play a minor role in Germany. In 2019, 8,919 megawatts of biomass were installed in Germany.

HYDROPOWER

Hydropower is certainly one of the oldest sources of energy and was already in use 5000 years ago—if not for the production of electricity but the operation of mills, for example. During the industrial revolution, hydropower was one of the most important pillars of energy generation. The first power-generating turbines were developed in the mid-19th century. In 1890 the first German hydropower plant, which was also the first AC power plant, went online in Bad Reichenhall. Hydropower tends to play a subordinate role in Germany and, with an installed capacity of 4,780 MW in 2019, is around 3% of gross electricity generation.

EEG

The predecessor of the EEG was the 1991 Electricity Feed-in Act, which allowed network operators to oblige decentralised operators to feed into the electricity grid. The law obliged network operators to buy electricity from producers with a minimum remuneration that corresponded to the average price that electricity generated two years earlier. The first EEG came into force in 2000. The central idea here was that renewable energies were given priority over other energy sources. Also, the remuneration rates for photovoltaics were increased significantly. While the rate in the feed-in electricity law was still around nine cents/kWh, it was between 48 and 50 cents/kWh in the EEG 2000. A sub-sidy cap of 300 MWp, which was recorded in the so-called 100,000 roof program (1999—the successor of the 1000 roof program), threatened to provoke a slump in the solar industry in 2004 since this had already been exceeded in 2003 with 350 MWp. The 2004 amendment, which was passed, for this reason, consisted primarily of an adjustment of the funding rates and a reduction in the funding of wind turbines.

The main goal of the EEG 2009 was to increase the share of renewable energies in the electricity mix to 30% by 2020. Also, the option appeared in the EEG 2009 for the first time that the network operator could regulate decentralised generation plants to minimise network fluctuations. The EEG 2012 is the basis for many elements of today's energy transition. This law introduced both direct marketing based on the market premium model and the flexibility premium. The first virtual power plants for bundling and marketing electricity from decentralised plants were created. The EEG 2014 focused more on the targeted development of renewable energies and defined fixed expansion paths for the individual energy sources. With this amendment, biogas plants disappeared of the agendas of legislators. Also, the idea of the call for funding was launched here, and mandatory direct marketing for new systems over 100 kW was introduced in the EEG 2014. To date, this was only valid from 750 kW.

In 2016, the last amendment to the EEG was made for the time being. Here the legislator made a fundamental change in the system. Previously, the feed-in tariff model was applied, but the Federal Government is now using tendering procedures, which have already been tested as a pilot project in the area of photovoltaic open-space systems in the EEG 2014.

WHY DO WE NEED ELECTRICITY AT HOME?

Your electricity consumption starts when you wake up to the sound of soft music or the shrill alarm of your clock. Do you have a cell phone? Then you must have filled it up with energy overnight through electricity from a socket or with batteries.. You take it off the charging cable and turn on the light. In winter, our lightbulbs have to perform well because the sun rises late and then sets early. Luckily, our lamps provide light—with what?

Next up is the bathroom. A cold shower is supposed to be healthy but is unfortunately only as pleasant as a visit to the dentist. In some households, to take a nice, warm shower the water is heated with electricity. Once you get out of the shower, you might plug in the hairdryer or electric razor. And to make your next visit to the dentist more enjoyable than a cold shower, maybe you brush your teeth with an electric toothbrush.

Power is now required in the kitchen. Your breakfast egg and your warm cocoa are cooked on the electric stove. The toaster also needs electricity to roast bread, and the coffee machine requires electricity to brew coffee. The oven, kettle, dishwasher, blender, and many other kitchen appliances also need electricity. The fridge, for example, continually consumes electricity to keep all food cool and fresh, and the freezer keeps your favorite ice cream from melting.

Had a great breakfast? Then off to school. The subway and S-Bahn trains run on energy, as do the headlights of cars and buses. The traffic light does not turn green without energy, and the street lamps shine our way with it.

In school, of course, the energy constantly flows. The lights are on in the classrooms. Projectors and other technology won't work without energy—not without electricity. Even clocks and the school bell need it.

Electricity is also constantly used in factories and offices. Escalators and elevators require electricity to take people to and from work. Phones and computers also need it. And in factories, machines use electricity to make your clothes, your toys, your school supplies, your crayons, your craft supplies, and almost everything else we need.

Usually, the baker bakes their goods in an electric oven, and the electricity in the supermarket ensures that your frozen pizza stays cold and the checkouts work.

Clearly, you brushed your teeth well this morning, and you probably don't have to go to the dentist. But if you do, the dentist needs electricity for his drill. Many other doctors also use devices that require electricity. It is particularly important in the hospital. Many electronic devices monitor patients' well-being and help restore them to health.

It can do much more: trains run it and take you from here to there at lightning speed. Even if airplanes fly with petrol, they can't fly without electricity. Airport runways are illuminated with it, and the technical devices onboard also need it.

An energetic day comes to an end. You may cuddle on the sofa with loved ones and watch TV. Or maybe you are still at the computer to play a little more. You probably already know: both devices need electricity. Before you go to sleep, put your cell phone back on the charging cable and set your alarm clock. Then the lights go out and sleep. After all, you will need much energy tomorrow.

CHAPTER 4: RELATIONSHIP BETWEEN VOLTAGE, CURRENT, RESISTANCE, AND POWER IN ELECTRICITY

Georg Ohm found that at constant temperature, the electrical current flowing through a fixed linear resistor is directly proportional to the applied voltage and inversely proportional to the resistance. This relationship between voltage, current, and resistance forms the basis of Ohm's law and is shown below.

OHM'S LAW

By knowing any two principles of the voltage, current, or resistance quantities, we can find the third missing value using Ohm's law. Ohm's law is used extensively in electronic formulas and calculations, so it is "very important to understand these formulas and remember them exactly."

To find the voltage (V):

[V = I x R] V (voltage) = I (ampere) x R (Ω)

To find the current (I):

$[I = V \div R]$ I (ampere) = V (voltage) $\div$ R (Ω)

To find the resistor (R):

$[R = V \div I]$ R (Ω) = V (voltage) $\div$ I (ampere)

It is sometimes easier to remember this Ohm law relationship using pictures. Here, the three quantities V, I, and R were superimposed into a triangle (called the Ohm law triangle), which supplies voltage with current and resistance beneath. This procedure signifies the definite location of each set within Ohm's law.

OHM'S LAW: THE MAGIC TRIANGLE

Implementing the above standard equation of Ohm's law gives the following combinations of the same equation:

Then, using Ohm's law, we can see that a 1V voltage applied to a 1Ω resistor will cause a 1A current to flow, and the larger the resistance value, the less current will flow for a given applied voltage. Every electrical device or component that obeys "Ohm's law," i.e. the current that flows through it, is proportional to the voltage across it (I α V), such as resistors or cables, is referred to as "ohmic," and devices that are not, like transistors or diodes, as "non-ohmic" devices.

Electrical force in circuits

Electrical force (P) in a circuit is the rate at which vitality is ingested or created inside a circuit. From a vitality source, a voltage produces or conveys power while the associated load is retaining it. Lights and warmers, for instance, assimilate electrical vitality and convert it into warmth, light, or both. The higher their worth or wattage, the greater power they are probably going to expand.

The amount image for power is P, and the result of voltage duplicated by the current unit of estimation is watt (W). Prefixes are utilized to mean the different multipliers or sub-multipliers of a watt, for example, milliwatts (mW = 10 - 3 W) or kilowatts (kW = 10 3 W).

On the off chance that one applies Ohm's law and replaces the estimations of V, I, and R, the recipe for electrical force can be found as follows:

The most effective method to discover the force (P) using:

[P = V x I] P(watts)

= V(volts) x I(amps)

[P = V2/R] P(watts) = V 2 (volts) / R (Ω)

[P = I 2 x R] P(watts)

= I 2(amps) R (Ω)

Once more, the three sizes were orchestrated in a triangle, this time the "power" triangle, with the power at the denominator and current and voltage at the nominator. This plan speaks to the real situation of any size inside Ohm's law.

THE POWER TRIANGLE

Again, the transformation of Ohm's law for electrical power gives us the following combinations of the same equation to find the different single quantities:

We see that there are three potential equations for figuring the electrical force in a circuit. On the off chance that the determined force is sure, (+ P) in the incentive for any recipe, the segment assimilates the force, i.e. it expends it. On the off chance that the determined force is negative, (− P), the part creates or produces power, and at the end of the day, it is a wellspring of electrical vitality, for example, batteries and generators.

ELECTRICAL POWER

Electrical segments have a "wattage" that shows the greatest rate at which the segment changes over electrical power into different types of vitality, for instance, a 1/4W resistor or a 100W light. Electrical devices convert one form of energy into another. For example, an electric motor will convert electrical energy into mechanical force, while an electrical generator will convert mechanical force into electrical energy. A lightbulb converts electrical energy into light and heat.

We now also know that the unit of power is called WATT, but some electrical devices such as electric motors have a power in the old measurement of "horsepower" or horsepower. The ratio between power and watt is as follows: 1PS = 746W . For example, a two-horsepower motor has an output of 1492 W, (2 x 746) or 1.5 kW.

OHM'S LAW AS A MATRIX TABLE

Ohm's Law

Find the voltage (V), current (I), resistance (R) and power (P) for the circuit shown below.

Voltage V

V = I (current) x R (resistance)

= 2 x 12Ω

= 24V

Current (I)

I = V/R

= 24/12Ω

= 2A

Resistance R

R= V/I

= 24/2

= 12 Ω

Power P

= V x I

= 24 x 2

= 48W

Power within a circuit is only available if BOTH voltage and current are available. For example, there is voltage at idle, but there is no current I = 0 (zero). Consequently, V x 0 is 0 , so that the power loss within the circuit must also be 0 . There is also current flow in the event of a short circuit, but there is no voltage $V = 0$, e.g. $0 \times I = 0$, so that the power loss within the circuit is again 0 .

Since electrical power is the product of $V \times I$, the power loss in a circuit is the same, whether the circuit contains high voltage and low current or low voltage and high current flow. In general, electrical energy is dissipated in the form of heat or mechanical work (such as motors). Energy can also dissipate in the form of radiation (lamps) or be stored (batteries).

ELECTRICAL ENERGY IN CIRCUITS

Electrical energy is the ability to do work. The unit of work or energy is joules (J). Electrical energy is the product of power multiplied by the time used. So if we know how much power is consumed in watts and how much time has passed in seconds, we can determine the total energy in watt-seconds. In other words, energy is equal to power x time and power = voltage x current. Therefore, the electrical power is dependent on the energy and the unit for electrical energy is watt-seconds or joules.

Electrical power can also be the rate at which energy is transferred. If a joule of work is absorbed or delivered at a constant rate of one second, the corresponding power corresponds to one watt, so that the power can be defined as "1 joule / sec = 1 watt." Then one can determine that a watt corresponds to a joule per second and the electrical power can be defined as working speed or energy transfer.

ELECTRICITY AND ENERGY TRIANGLE

We previously predicted that electrical energy is defined as watts/second or joules. Moreover, because electrical energy is measured in joules, it can have a large value when calculating the energy consumption of a component. For example, if a 100-watt light bulb is left on for 24 hours, the energy consumption is 8,640,000 joules (100 W x 86,400 seconds), so that prefixes such as kilojoules (kJ = 10 3 J) or megajoules (MJ = 10 6 J) are used and in this simple example the energy consumption is 8.64MJ (Mega-Joule).

When dealing with joules, kilojoules, or megajoules to express electrical energy, math can end up with some large numbers and many zeros, making it much easier to express electrical energy consumed in kilowatt hours.

If the electrical power consumed or generated can be measured in watts or kilowatts, i.e. thousands of watts and time in hours rather than seconds, then the unit of electrical energy is kilowatt hours (kWh). Then our 100 watt light bulb consumes 2,400 watt hours or 2.4 kWh, which is much easier to comprehend than 8,640,000 joules.

The amount of electricity consumed by a device with an output of 1000 watts per hour is 1 kWh and is generally referred to as the "unit of electricity." That is what the consumption meter measures and what we as consumers buy from our electricity suppliers.

Kilowatt hours are the standard units of energy used by the electricity meter in our homes to calculate the amount of electrical energy that we use and thus how much we pay. Consequently, if you switch on an electric light or a heating element with a power of 1000 watts and keep it switched on for 1 hour, you have consumed 1 kWh of electricity. If you turned on two electric lights, each with 1000 watt elements, for half an hour, the total consumption would be the same amount of electricity—1kWh.

The consumption of 1000 watts per hour therefore consumes the same power as 2000 watts (twice as much) per half hour (half the time). For a 100 watt light bulb to consume 1 kWh or a unit of electrical energy, it would then have to be switched on for a total of 10 hours (10 x 100 = 1000 = 1kWh).

Now that we know the relationship between voltage, current, and resistance in a circuit, we'll look at the standard units used in electrical engineering to calculate them and see that each value is represented either by a multiple or part of the standard unit can be displayed.

CHAPTER 5: OPTOELECTRONICS, PLASMONS (PHYSICS)

The term optoelectronics (sometimes also called optronics or optronics) originated from the combination of optics and semiconductor electronics and broadly encompasses all products and processes that enable the conversion of electronically generated data and energies into light emission and vice versa.

The background is, e.g., B. the attempt to combine the advantages of electronic data processing and processing with the advantages of fast and electromagnetic and electrostatically undisturbed broadband transmission property of light. At the same time, this also includes the conversion of electrical energy into light and vice versa based on electronic semiconductor technology, whereby the generated light can either spread in open space or solid translucent media (optical fibers, such as glass fiber cables) or, as in optical storage technology, can also be used to store electronically generated data.

Optoelectronics have become an integral part of everyday life because they include components such as lasers, screens, computers, optical storage, and data carriers.

OPTOELECTRONIC COMPONENTS

Optoelectronic components act as an interface between electrical and optical components or devices that contain such components. This mostly (but not exclusively) means microelectronic components that function based on semiconductors.

The components of optoelectronics can be divided into actuators (transmitters) and detectors (receivers). Optoelectronic actuators are semiconductor components that generate light from electricity, e.g. laser and light-emitting diodes. The emission spectrum can be visible or invisible (UV or infrared). Optoelectronic detectors are the reversing components of the actuators, e.g. photoresistor, a photodiode (also solar cell) and phototransistor. Light sensors can also be built as an integrated circuit, e.g. as a CCD sensor. Also, photomultipliers are counted among optoelectronics. If the actuator and detector operate in a system, this results in an optical sensor, a so-called Opto sensor. The specialist area is also referred to as Opto sensor technology. The simple combination of an actuator and a detector in one component is called an optocoupler.

In addition to these, other components are required for the transmission, amplification, or modulation of signals. Optical signals can be transmitted through free space or in conjunction with waveguides and optical circuits (see integrated optics). Optical modulators are components that impress (modulate) light on a defining characteristic. This can be a temporal or spatial amplitude or phase variation, for example. These include optical amplifiers, optoelectronic multiplexers, and magnetostrictive optical micro reflectors.

PLASMONS

The plasmon was first proposed in 1952 by David Pines and David Bohm and appeared to emerge from a Hamiltonian for the long-extend electron-electron connections. Since plasmons are the quantization of traditional plasma motions, the greater part of their properties can be discovered from Maxwell's conditions. In material science, a plasmon is a quantum of plasma swaying. Similarly, as light (an optical swaying) comprises photons, the plasma wavering comprises plasmons. The plasmon can be considered a quasiparticle since it emerges from the quantization of plasma motions, much like phonons quantify mechanical vibrations. Along these lines, plasmons are aggregate (a discrete number) motions of the free electron gas thickness. For instance, at optical frequencies, plasmons can couple with a photon to make another quasiparticle called a plasmon polariton.

Plasmons can be depicted in the old-style picture as a wavering of electron thickness for the fixed positive particles in the metal. To envision a plasma swaying, envision a 3D square of metal put in an outside electric field, highlighting the right. Electrons will move to one side (revealing positive particles on the correct side) until they drop the field inside the metal. On the off chance that the electric field is evacuated, the electrons move to one side, repulsed, and pulled into the positive particles left uncovered on the correct side. They waver to and fro at the plasma recurrence until the energy is lost in an obstruction or dampening. Plasmons are a quantization of this sort of swaying.

Plasmons assume a huge job in the optical properties of metals and semiconductors. A material reflects frequencies of light underneath the plasma recurrence because of the electrons in the material screen the light's electric field. A material transmits light of frequencies over the plasma recurrence because the electrons in the material can't react sufficiently quickly to screen it. In many metals, the plasma recurrence is bright, making them glossy (intelligent) in the obvious range. A few metals, such as copper and gold, have electronic interband advances in the obvious range, whereby explicit light energies (hues) are retained, yielding their unmistakable shading. In semiconductors, the valence electron plasmon recurrence is, for the most part, in the profound bright. Their electronic interband advances are in the obvious range, whereby explicit light energies (hues) are ingested, yielding their particular shading, which is the reason they are intelligent. It has been demonstrated that the plasmon recurrence may happen in the mid-infrared and close infrared locale when semiconductors are as nanoparticles with overwhelming doping.

Surface plasmons

Surface plasmons are those plasmons that are restricted to surfaces and that connect firmly with light, bringing about a polariton. They happen at the interface of a material displaying a positive, genuine piece of their relative permittivity, such as dielectric consistent (e.g. vacuum, air, glass, and different dielectrics), and a material whose genuine piece of permittivity is negative at the given recurrence of light, normally a metal or intensely doped semiconductors. Notwithstanding inverse indication of the genuine piece of the permittivity, the size of the genuine piece of the permittivity in the negative permittivity district ought to normally be bigger than the extent of the permittivity in the positive permittivity area. In any case, the light isn't bound to the surface (for example the surface plasmons don't exist) as stated in the renowned book by Heinz Raether. At obvious frequencies of light, such as 632.8 nm frequency from a He-Ne laser, interfaces supporting surface plasmons are frequently framed by metals like silver or gold (negative genuine part permittivity) in contact with dielectrics, such as air or silicon dioxide. The specific selection of materials can drastically affect the level of light imprisonment and proliferation separation because of misfortunes. Surface plasmons can likewise exist on interfaces other than level surfaces, such as particles, rectangular strips, v-sections, chambers, and different structures. Numerous structures have been explored because of surface plasmons' capacity to keep light beneath the diffraction furthest reaches of light.

One straightforward structure that was examined was a multilayer arrangement of copper and nickel. Mladenovic et al. report the utilization of the multilayers as though it's one plasmonic material. The copper oxide is forestalled with the expansion of the nickel layers. It is a simple way to coordinate plasmonics to utilize copper as the plasmonic material, since it is the most widely recognized decision for metallic plating alongside nickel. The multilayers fill in as a diffractive grinding for the occurrence light. Up to 40% of transmission can be accomplished at typical occurrence with the multilayer framework relying upon the proportion of copper to nickel. In this manner, the utilization of effectively famous metals in a multilayer structure can answer plasmonic coordination.

Surface plasmons can assume a job in surface-improved Raman spectroscopy and clarify oddities in diffraction from metal gratings (Wood's peculiarity), among other things. Surface plasmon reverberation is utilized by natural chemists to consider the components and kinetics of ligands authoritative to receptors (for example, a substrate authoritative to a chemical). Multi-parametric surface plasmon vibration can be utilized not exclusively to quantify atomic connections but also to nanolayer properties or basic changes in the adsorbed particles, polymer layers, or graphene.

Surface plasmons may likewise be seen in the X-beam emanation spectra of metals. A scattering connection for surface plasmons in the X-beam discharge spectra of metals has been inferred (Harsh and Agarwal).

As of late, more surface plasmons have been utilized to control shades of materials. This is conceivable since controlling the molecule's shape and size decides the kinds of surface plasmons that can be coupled into and spread across it. Consequently, this controls the association of light with the surface. These impacts are delineated by the memorable recolored glass which decorates medieval houses of prayer. Metal nanoparticles deliver some recolored glass hues of a fixed size which collaborate with the optical field to give the glass a dynamic red shading. In present-day science, these impacts have been designed for both obvious light and microwave radiation. Much exploration goes on first in the microwave run because material surfaces and tests can be delivered precisely at this frequency. After all, the examples will, in general, be on the request for a couple of centimeters. The creation of optical range surface plasmon impacts includes making surfaces that have highlights <400 nm. This is considerably more troublesome, yet has become conceivable to accomplish in any solid or accessible manner.

As of late, graphene has also appeared to oblige surface plasmons, as watched through close to handle infrared optical microscopy methods and infrared spectroscopy. Expected uses of graphene plasmonics primarily tended to the terahertz to midinfrared frequencies, such as optical modulators, photodetectors, and biosensors.

Plasmon-Soliton

Plasmon-Soliton scientifically alludes to the crossbreed arrangement of nonlinear plentifulness conditions, such as for a metal-nonlinear media considering both the plasmon mode and singular arrangement. A soliplasmon reverberation is then again considered as a quasiparticle consolidating the surface plasmon mode with spatial soliton because of a resounding collaboration. To accomplish one-dimensional singular proliferation in a plasmonic waveguide while the surface plasmons ought to be limited at the interface, the sidelong dissemination of the documented encompass ought to likewise be unaltered.

A graphene-based waveguide is a reasonable stage for supporting half breed plasmon-solitons because of the enormous compelling zone and immense nonlinearity. For instance, the proliferation of lone waves in a graphene-dielectric heterostructure may show up as higher request solitons or discrete solitons coming about because of the opposition among diffraction and nonlinearity.

CHAPTER 6: SOLAR PANEL CONSTRUCTION

Solar panel technology is progressing quickly with more noteworthy productivity and lower costs bringing about a colossal increment sought after. Regardless of solar panel's progression, fundamental solar panel construction hasn't changed as much. Most solar panels are as yet comprised of a progression of crystalline silicon cells sandwiched between a front glass plate and a back polymer plastic back-sheet upheld inside an aluminum outline.

Once introduced, solar panels are exposed to extreme conditions all through their 25+year lifespan. Outrageous varieties in temperature, stickiness, wind, and UV radiation can put tremendous weight on a solar panel. Luckily, most panels are designed to withstand the outrageous climate. In any case, a few panels can even now fall flat in a few different ways, including water entrance, cell smaller scale cracks, and likely actuated corruption or PID. This is the reason it is imperative that solar panels are produced utilising only the best segments. We feature the main makers utilising the best materials and testing the most noteworthy industry gauges. See the manual for choosing the greatest solar panels.

HOW ARE SOLAR CELLS MADE?

Solar photovoltaic PV cells are made utilising crystalline silicon wafers, which are like the wafers used to make PC processors. They can either be polycrystalline or monocrystalline silicon wafers and are delivered utilising a few diverse assembling techniques. The most effective sort is monocrystalline (mono), which is fabricated utilising the notable Czochralski process. This procedure is more vitality-concentrated compared to polycrystalline (poly) and accordingly increasingly costly to deliver.

Polycrystalline wafers are marginally less effective and are made utilising a few cleansing procedures followed by a less complex, lower cost strategy. All the more, as of late, cast monocrystalline or cast mono cells have picked up in ubiquity. The explanation is the lower-cost throwing process used to make cast mono cells like the procedure utilised for polycrystalline silicon cells. Be that as it may, cast-mono wafers are not perfectly productive and unadulterated mono wafers made utilising the Czochralski procedure.

- Monocrystalline silicon cells - Highest effectiveness and greatest expense
- Cast mono silicon cells - High effectiveness and lower cost
- Polycrystalline silicon cells - Lower effectiveness and cheapest

ASSEMBLING CRYSTALLINE SILICON CELLS

In the first place, silicon is separated from the sand. The sand utilised is known as silica sand or silicon dioxide and is typically produced using squashed quartz rock. The silica sand should then be filtered utilising a procedure called Carbon Arc Welding (CAW), which separates the undesirable oxygen and results in 99% pure silica. The silica is then additionally handled to become as near 100% pure silicon. The final product is pure polycrystalline silicon, which can be doped with the following measures of either boron or phosphorous to turn out to be either N-type or P-type silicon. Next to this stage, polycrystalline silicon can be liquefied and thrown into large rectangular squares and meagerly cut utilising a jewel wire slicing strategy to create the polycrystalline or multi-crystalline wafers.

To produce the more proficient monocrystalline wafer or cells, the doped silicon can be made into a pure strong precious stone ingot utilising the Czochralski procedure. This procedure includes softening the polycrystalline silicon under high tension and temperature to gradually grow a single monocrystalline precious stone known as an ingot.

STEPS TO MAKE MONOCRYSTALLINE PV CELLS

- Silica sand is cleaned to 99% silicon utilising the CAW procedure
- The 99% silicon is additionally refined near 100% unadulterated silicon
- The silicon is doped with phosphorous or boron (P or N-type)
- The doped silicon is softened and separated into a solitary ingot
- The enormous round ingot is jewel wire-cut into meager square wafers

- The flimsy wafers are covered with a meagre layer of either P- or N-type to shape the PN-intersection
- Fine metallic fingers are screen imprinted onto the cells
- Level lace busbars (as appeared) or flimsy wire (MBB) busbars are included

HOW ARE SOLAR PANELS MADE?

Solar panels are made utilising the six primary parts portrayed underneath and collected in cutting edge fabricating offices with extraordinary precision. We will concentrate on panels made utilising silicon crystalline solar cells, which are the most well-known and most noteworthy performing solar technology accessible today. There are other solar PV advances accessible, such as dainty film and screen printed cells; however, we won't talk about these, as they have restricted use or are still being developed.

SOLAR PHOTOVOLTAIC CELLS

- Toughened Glass—3 to 4mm thick
- Expelled aluminum outline
- Exemplification—EVA film layers

- Polymer back sheet
- Intersection box—diodes and connectors

Many notable solar panel producers are vertically incorporated, which implies the one organisation supplies and makes all the fundamental segments, including the silicon ingots and wafers used to make the solar PV cells. Many panel makers collect solar panels utilising remotely sourced parts, including cells, polymer back sheet, and embodiment EVA material. These makers can be increasingly particular about which segments they choose. However, they don't generally have authority over the nature of the items, so they ought to be certain they utilise the best providers accessible.

1.   SOLAR PV CELLS

Solar photovoltaic cells or PV cells convert daylight straightforwardly into DC electrical vitality. The cell type dictates the solar panel execution, and qualities of the silicon utilised, with the two fundamental sorts being polycrystalline and monocrystalline silicon. The base of the cell can be made utilising various added substances to make either positive p-type silicon or negative n-type silicon. There are a few distinctive cell sizes and setups accessible, which offer various degrees of proficiency and execution, including half-cut or split cells, multi-busbar (MBB) cells, and shingled cells utilising slight covering wafer strips. For progressively nitty-gritty data on the various cells and solar panels types, see the total solar PV cell technology survey.

Most private solar panels contain 60 mono or polycrystalline cells connected through busbars in arrangement to create a voltage between 30 and 40 V depending upon the kind of cell used. Bigger solar panels for business frameworks and utility-scale solar ranches contain 72 to 96 cells, which works at a higher voltage. The electrical links which interconnect the cells are known as busbars and permit the flow to course through all the cells in a circuit.

2.   GLASS

The front glass sheet shields the PV cells from the climate—from hail or airborne flotsam and jetsam. The glass is regularly high-quality safety glass, which is 3 to 4 mm thick and structured to oppose mechanical burdens and extraordinary temperature changes. The IEC least standard effect test requires solar panels to withstand an effect of hailstones of 1 inch (25 mm) breadth venturing out up to 60 mph (27 m/s). In the case of a mishap or extreme effect, treated glass is additionally a lot more secure than standard glass, as it breaks into minuscule pieces as opposed to sharp, rough segments.

To improve the proficiency and execution of the panel, high transmissive glass is utilised by most producers with low iron substance and an enemy of intelligent covering on the backside to lessen misfortunes and improve light transmission.

3.   ALUMINIUM FRAME

The aluminium outline assumes a basic job by both securing the edge of the overlay area, lodging the cells, and giving a strong structure to mount the solar panel in position. The expelled aluminium segments are intended to be amazingly lightweight, solid, and ready to withstand extraordinary pressure and to stack from high wind and outer powers.

The aluminum casing can be silver or anodised dark, and relying upon the panel producer, the corner areas can either be squeezed or clipped together, giving various degrees of solidarity and solidness.

4.   EVA FILM

EVA means 'ethylene vinyl acetic acid derivation,' which is an uncommonly structured polymer profoundly straightforward (plastic) layer used to exemplify the cells and hold them in position during produce. The EVA material must be incredibly strong and adaptable to withstand extreme temperature and moisture, and the material has a significant influence in the drawn-out exhibition by forestalling dampness and earth entrance.

The overlay on either side of the PV cells gives some stun assimilation and shields the cells and interconnecting wires from vibrations and unexpected effects from hailstones and such. A great EVA film with considerable 'cross-connecting' can be the distinguished between a long life or a panel disappointment because of water entrance. During production, the cells are first built with the EVA before being gathered inside the glass and back sheet.

5.   BACKSHEET

The back sheet is the back-most layer of basic solar panels, which constitutes a dampness obstruction and last outside skin to give both mechanical insurance and electrical protection. The back sheet material is made of different polymers or plastics, including PP, PET, and PVF, which offer various degrees of insurance, warm steadiness, and long haul UV opposition. The back-sheet layer is normally white in shading but, on the other hand, is accessible as clear or dark depending upon the maker and module.

Some panels, such as bifacial and frameless panels, utilise a back glass panel rather than a polymer back sheet. The backside glass is more sturdy and enduring than most back-sheet materials. Thus, a few makers offer a multi-year execution guarantee on double-glass panels.

6.   JUNCTION BOX AND CONNECTORS

The intersection box is a little climate confirmation fenced-in area situated on the backside of the panel. It is intended to join the links required to interconnect the panels safely. The intersection box is significant, as it is the main issue where all the cells' sets interconnect and should be shielded from dampness and soil.

SIDESTEP DIODES

The intersection box likewise houses the detour diodes, which are expected to forestall back current, which happens when a few cells are concealed or filthy. Diodes permit electrical current to stream one way, and a common 60 cell panel has three columns of 20 PV cells. There are three detour diodes, one for forestalling reverse current to every one of the three arrangements of cells. Unfortunately, sidestep diodes can come up short after some time and maybe supplanted. Hence, the front of the intersection box is typically ready to be evacuated for overhauling, albeit numerous cutting edge solar panels presently utilise further developed durable diodes and non-functional intersection boxes.

SOLAR MC4 CONNECTORS

Practically all solar panels utilise unique climate-safe attachments called MC4 connectors. The term MC4 represents a multi-contact 4 mm measurement connector. Because of the outrageous climate conditions, the connectors must be exceptionally vigorous, secure, UV safe, and keep a decent association with insignificant opposition at both low and high voltages up to 1000 V.

The connectors are intended to be utilised with the standard 4 mm or 6 mm twofold protected solar DC link with a tinned copper multi-strand centre for the least opposition. To accurately collect the connectors, an uncommon pleating device is utilised to create the multi-strand link to the internal terminal, which is then embedded and snapped into the MC4 lodging.

There are a few assortments of MC4 connectors that may appear to be comparable yet don't generally fit together safely. A similar sort and make of connector should consistently be utilised to decrease potential water entrance or attachment disappointment, which can bring about arcing and even fire. The current MC4 appeared above. Cutting edge MC4-EVO-2 connectors (not appeared) are both made by Staubli and the main disparate looking connectors, which are permitted to be utilised together.

SOLAR PANEL ASSEMBLY AND MANUFACTURING

Solar panels are gathered in cutting-edge producing offices with mechanised, automated hardware and sensors to situate the segments decisively with extreme precision. The assembling plants must be incredibly spotless and controlled to forestall any sullying during getting together.

All through the assembling procedure, the panels and cells are checked and investigated utilising progressed optical/imaging sensors to guarantee all the parts are found accurately and the cells wafers, which are fragile, are not harmed or broken. Contingent upon the maker, the last panel get-together is completely checked to utilise various tests, including electroluminescent (EL) or blaze testing, to distinguish any imperfections in the cells which could prompt disappointment once presented to daylight and high temperatures for a long time.

SUSTAINABILITY

Daylight or solar vitality is a free wellspring of a sustainable power source that can never be exhausted. Then again, petroleum derivatives are a limited asset that transmits ozone harming substances and different particulates during extraction, handling, and burning. During examination, solar panels don't create any discharges while being used, yet they are produced using a few distinct materials which require various degrees of assets and vitality. The vitality used to remove the crude materials and to assemble an item is known as the 'typified vitality.' The measure of time it takes for an item to reimburse the typified vitality is estimated in years. This is known as the absolute vitality compensation time (EPBT).

A run of the mill silicon crystalline solar panel will create enough vitality to reimburse the epitomised vitality inside two to three years of establishment. In any case, as panel effectiveness has expanded, the compensation time has decreased to under two years in numerous regions with high normal solar radiation.

Current crystalline silicon solar panels produce enough vitality to reimburse the encapsulated vitality inside two to three years of establishment. Numerous definite examinations and life-cycle investigation bolster this. In any case, a large number of the examinations are currently obsolete as solar PV cell productivity has expanded from 15% to 20% (a 35% expansion) throughout the most recent couple of years, and recompense time is assessed to be as low as on and a half years. Considering a run of the mill solar panel will last 20 to 30 years, it will effortlessly reimburse the exemplified vitality on numerous occasions over and counterbalance a large number of huge amounts of discharges.

STEPS TO MAKING SOALR PANEL FOR PERSONAL HOME USE

Solar energy is a sustainable wellspring of energy that benefits you as well as the earth. Through the exertion you put into making a custom-made solar board, you can help forestall natural contamination by lessening petroleum product utilisation. What's shockingly better is that you'll get a good deal on your electric bill. To fabricate your solar board, you'll have to gather the pieces, interface the cells, assemble a board box, wire the boards, seal the case, and lastly, mount your finished solar board.

CHAPTER 7: SOLAR PANEL CONSTRUCTION

STAGE 1: ASSEMBLING THE PIECES

BUY THE CELLS

There are a couple of various solar cells to purchase, and most great options are either made in the United States, China, or Japan. Nonetheless, the best expense-to-proficiency choice is probably polycrystalline cells. The quantity of cells you should purchase depends upon the measure of energy you're hoping to obtain. The specs ought to be recorded when you buy the cells.

- Make sure to purchase additional items. These cells are incredibly delicate.
- Cells can be effectively purchased online through sites like eBay; however, you may buy some from your nearby home improvement shop.

- It might be important to clear wax off the cells if the producer ships them in wax. To do this, dunk them in hot, though not boiling, water.
- Each cell shouldn't cost more than $1.30 per watt.

MEASURE AND CUT A SPONSORSHIP BOARD

You will require a dainty board made out of a non-conductive material, such as glass, plastic, or wood, to append the cells. Determine the solar panel plan, and measure and slice a board to that size. Leave an additional inch or two on the ends for the wires that interface the lines. Wood is a basic sponsorship material to pick since it's simpler to penetrate, which you'll have to do in order for the cell wires to go through.

MEASURE AND CUT THE ENTIRETY OF YOUR SELECTING WIRE

The moment you take a gander at your polycrystalline cells, you'll see numerous little lines going one way and two bigger lines going the other way. You should associate a selecting wire with running down the two bigger lines and interfacing with the cell's back in the cluster.

Measure the length of that bigger line, double the length, and cut two pieces for every cell.

MOTION IN THE WORK TERRITORY

Utilising a motion pen, run two to three lines of transition down every cell strip's length, or gathering of three squares. Make a point to do this on the back of the cells. This will shield the warmth of the patching from causing oxidation.

BIND THE SELECTING

Utilise a binding iron to place a slim layer of patch onto the rear of the cell strips. This step is unnecessary if you buy pre-fastened selecting, which is regularly a superior choice since it cuts time, warms the cells just a single time, and squanders less weld. Be that as it may, it's increasingly costly.

BOND THE WIRE TO THE CELLS

Warm the main portion of a bit of selecting wire with a welding iron. Bond the wire's end to a cell. Rehash this holding procedure for every cell.

STEP 2: ASSOCIATING THE CELLS

PASTE THE CELLS TO THE BOARD

Put a limited quantity of paste at the back focal point of the cells and press them into place. The selecting wire should run in a solitary, straight line. Ensure the selecting wire closes are coming up between the cells and are allowed to move, with simply the two pieces standing up between every cell. Make sure that one column runs toward the path inverse to the one close to it so that the selecting wire stands out toward the finish of one line and on the contrary side of the following. You should plan to place the cells in long lines with less lines. For instance, three columns, each comprising of 12 cells, put long side to long side. Remember to leave an additional inch (2.5 cm) at the two closures of the load up.

PATCH THE CELLS TOGETHER

Apply a transition to the length of the two thick lines (contact cushions) on every cell. Take the free areas of selecting wire and bind them to the cushions' whole length. The selecting wire associated with the rear of one cell ought to interface with the following cell's front for each situation.

INTERFACE THE PRIMARY COLUMN UTILISING TRANSPORT WIRE

Toward the start of the primary line, weld the selecting wire to the front of the main cell. The selecting wire ought to be about an inch (2.5 cm) longer than expected to cover the lines and reach out towards the board's additional hole. Patch those two wires along with a transport wire piece that is a similar size as the separation between the thick lines of the cell.

ASSOCIATE THE SUBSEQUENT COLUMN

Interface the finish of the main column to the start of the second with a long transport wire piece that stretches out between the wire at the edge of the board and the wire that is the furthest away in the following line. You should set up the primary cell of the second column with extra selecting wire, as you did with the first. Connect each of the four wires to this transport wire.

KEEP INTERFACING THE REMAINDER OF THE COLUMNS

Keep interfacing the columns with the long transport wires until you arrive at the end, associating it with a short transport wire once more.

STEP 3: BUILDING YOUR PANEL BOX

MEASURE YOUR CELL BOARD

Measure the space taken up by the board on which you have put your cells. You will require the container to be huge. Include 1 inch (2.5 cm) on each side to permit space for the case. If there won't be a free 1 inch by 1 inch (2.5 cm x 2.5 cm) square spot at each corner in the wake of including the board, leave space for this. Try to have adequate space for the transport wires toward the end also.

CUT THE LEVEL BACK

Slice a bit of pressed wood to match what you estimated in advance, in addition to the space for the crate sides. You can utilise a table saw or jigsaw, contingent upon what you have accessible.

STRUCTURE THE SIDES

Measure two 1 inch by 2 inches (2.5 cm x 5 cm) bits of a non-conductive board on the long sides of the case's base. Then measure two progressively 1 inch by 2 inches (2.5 cm x 5 cm) boards to fit between these long pieces, finishing the case. Cut these pieces you have estimated and secure them together utilising deck screws and butt joints. The sides mustn't be too tall since they may conceal the cells when the sun originates from a sharp point.

APPEND THE SIDES

Utilising deck screws, screw through the head of the sides and into the base to tie down the gatherings to the case's base. The quantity of screws you use per side will depend upon the sides' length; however, regardless of the length, you shouldn't utilise less than three.

PAINT THE CRATE

You can paint the crate whatever color you like. Consider utilising white or another light hues, since this will keep the container cooler and cells perform better when they are cool. Your board will last longer if you use paint intended for open-air use. This sort of paint will help shield the wood from the climate.

APPEND THE SOLAR UNIT TO THE CASE

Paste the solar unit to the finished box. Ensure that it is secure and that the cells are looking up to take in daylight. There ought to be two openings on the board to finish the transport wire.

STEP 4: WIRING YOUR PANEL

ASSOCIATE THE LAST TRANSPORT WIRE TO A DIODE

Get a diode somewhat greater than your board's amperage and associate it to the transport wire, protecting it with some silicone. The light-shaded end of the diode ought to point towards where the negative end of the battery goes. The opposite end ought to be wired to the negative finish of your board. This keeps energy from going back through the solar board from the battery when it isn't charging.

ASSOCIATE DIFFERENT WIRES

Associate a dark wire to the diode and run it to a terminal square, which you should mount on the crate's side. Then associate a white wire from the short transport wire on the contrary side to the terminal square.

INTERFACE YOUR BOARD TO A CHARGE CONTROLLER

Buy a charge controller and interface the board to the controller to associate the positive and negative effects. Run the wires from the terminal square to the charge controller, utilising shading coded wire to monitor the charges. If utilising more than one board, you should associate the entirety of the positive and negative wires utilising rings, to ensure you end up with two wires. Associate the charge controller to your batteries. Purchase batteries that will work with the size of the boards you assembled. Interface the charge controller to the batteries as indicated by the maker's directions.

UTILISE THE BATTERIES

When you have the batteries associated and charged from the board or boards, you can run your hardware off the batteries, contingent upon the measure of the intensity you need.

STEP 5: FIXING THE BOX

GET A BIT OF PLEXIGLASS

Buy a bit of plexiglass that is sliced to fit inside the container that you made for your board. You can get this from a forte shop or your nearby home improvement shop. Make sure you get plexiglass and not glass, as glass is liable to break or chip.

CONNECT SQUARE STOPS FOR THE GLASS

Cut 1 inch by 1 inch (2.5 cm x 2.5 cm) squares of wood to fit into the corners. These ought to be sufficiently high to fit over the terminal square yet low enough to fit underneath the case's lip. Paste these stops into place utilising wood stick.

SUPPLEMENT YOUR PLEXIGLASS

Fit the plexiglass onto the crate with the goal that the glass lays on the head of the squares. Utilising the suitable screws and a drill, cautiously screw the plexiglass into the squares.

SEAL THE CASE

Utilise a silicone sealant to seal the edges of the case. Additionally, seal any holes you can discover so that the container is as watertight as possible. Utilise the maker's guidelines to apply the sealant appropriately.

STEP 6: MOUNTING YOUR PANELS

Mount your boards on a truck. One option is to manufacture and mount your boards on a truck. This fix put the board at a point yet still permit you to alter which course the board countenances to adjust the amount of sun it gets in a day. Expect to alter the board two to three times each day.

MOUNT YOUR BOARDS ON YOUR ROOFTOP

This is a famous method to mount the boards, since utilizing it results in the solar panels getting, in general, the most daylight and staying off the beaten path. Be that as it may, the edge should be reliable with the sun's way and your pinnacle load time. This will restrict you to only getting full presentation at specific times. This choice is ideal on the off chance that you have numerous boards and next to no ground space to put them on.

MOUNT YOUR BOARDS ON A SATELLITE STAND

The stands generally used to mount satellite dishes can likewise be utilised to mount solar boards. They can even be modified to move with the sun. Be that as it may, this choice will work best if you only have few solar boards.

ARE SOLAR PANELS TOXIC?

Regardless of the enormous measure of data coursing about solar panels being harmful, present-day crystalline silicon solar panels contain no poisonous materials. The cases of 'poisonous solar panels' originated from the generally out of date flimsy film (Cadmium telluride - CdTe) solar panels, which contained measures of cadmium and telluride. Nonetheless, unless these (generally uncommon) panels are separated into pieces, the following measure of cadmium is contained inside the EVA layers and can't filter out.

Current crystalline silicon solar panels contain a follow measure of lead in the weld utilised for the cell interconnections. In any case, even the utilisation of bind is starting to be exchanged with the new busbar pressure joining methods and conductive glue materials. Its significant bind is utilised in countless electrical gadgets and apparatuses. There are unquestionably, increasingly harmful components utilised in shopper electronic gadgets, cell phones, PCs, and TVs, which is the reason electronic waste or E-squander is an enormous worldwide issue.

Generally, 98% of the solar panels introduced far and wide today are of the crystalline silicon assortment and don't contain cadmium or telluride. Solar panels are kind, and when harmed, the cells don't cause any sullying, as the cells are epitomised and contain no promptly solvent materials. As similar to other apparatuses, solar panels should be gathered and reused until the end of its life, which we examine below.

REUSING SOLAR PANELS

Since most solar panels introduced in the course of the most recent 20 years are being used, there is not an incredible volume of solar waste. It maybe throughout the following 10-20 years that many frameworks will arrive at the finish of life (EOF), and there would be a subsequent increase in the volume of solar-related waste, which should be reused. Solar panel reusing is a rising industry because of the effortlessly reused materials, including the aluminium casings and mounting frameworks. Most solar panel makers are pushing to be progressively sustainable. They are presently part of the not-revenue driven PV Cycle association—"PV CYCLE offers individuals and waste holders better access to reclaim and guarantee reusing rates over the business principles."

In Europe, the French waste administration organisation Veolia has opened the main solar panel reusing office in southern France, which can recuperate and reuse 95% of the materials.

CONCLUSION

The right way to live in this world is to make sure that there is continuity in our everyday activities, many of which are connected with electricity. But for electricity to reach the rural areas, it will take time, though you now have the power and knowledge to build one. Congratulations! Now you can now install solar panels on your property and enjoy uninterrupted electricity to go about your everyday life.